TECHNOLOGY

PROCESS.

A NATURAL AND EVOLUTIONARY PROCESS.

F. A. Nsiah

PART I

Technology Series

DEDICATION

This study notebook is dedicated to students,

teachers, friends and all.

Preface

Technology is normally pursued only as a practical subject in schools of engineering. Technology is move into the information society or age. Human is specialized in one respect or another. The ability to think abstractly, which allows the species to be at once specific and general. Technology is simply the applied science and the scientific discoveries that can easily converted into applications. Technology is by abstracting of specie characteristics from the natural process of organism. The abstraction of specie characteristics evolve both externally and rapidly into artifacts. Artifacts are artificial construction that mimics natural principles to do job rather than physical changes. The depth of technology in this book is based on evolutionary and natural processes that are reviewed on introduction to technology and main topics namely; cat control knowledge, cat simulation model, pounding task, wusair simulation and wusair graph. These are generalized by specifics only in understanding the construct of technology. It reviews technology on the principles of natural and evolutionary processes- Adaptability, Specialization, Survival of Fittest and Natural Selection. The approach is both natural and evolutionary in some physical sense to think to put the concept of technology in perspective and to adapt. The society, culture and technology are made clear in our understanding of the influences of human beings in our world ecosystem. Cat control[11] knowledge deals with the philosophical, psychological, physical views of cat designs and the organizational behavior of the cat that makes it intelligent. Cat simulation[12] models the cause-effect relationship of cat as designed in its environment. The organizational behavior of cat that makes intelligent and interact in a way of it's environment is a simulation. A pounding task[3, 4, 5] is based on the pistol-mortar activity of traditional Ashantians. It is the same concept in expressing the task oriented design but

the focus is different. These are repetitions for

clarity of thought. The wusair simulation of task-

orientation is designed to model the activity of

pistol-mortar. *Hand-side Architecture[n.s.] is* an

automaton[7, 8] is used in modeling the idea of

Von Neumann's machine[6] today called computer.

The technology of the computer is based on

Neumann's idea of self-reproduction[6]. In this

sense, the author's thought process is to model

solutions based on the computer technology by

only looking at the self-reproduction of the natural

process. By this and only this is the main means to

reproduce a self-machine from our natural and

evolutionary process. A pen-analogy is a natural-

evolutionary means to write as to save information

on paper. By this analogy, all things done by the

pen in a hand can be reproduced by a computer

technology. The thought process of what is even

representable on computing machinery is a mimic

of the pen-analogy. The technology for pen-

analogy is a machine revelation. Von Neumann's

complexity and mathematical conditions of

machine reproduction are will be deter-mined in

the next book series, n.s. It is about the machine

processing of pen-analogy with finite automation

and state transition tables. The automatic pen

controller is a simulation model based on the state

transition tables produced from finite automaton.

The computational model[7, 8] of finite automaton

is used to grasp the methodology of modeling a

pen control technology and will appear in the next

series. The transition function of a writing finite

automaton is given as the automatic pen controller.

The art of technology[10] in the specification of

real things as a notation is method of study and

investigation. This draws on the art specie by

explaining the production, thing (self), body and

skills properties of the real thing in a real world.

Author: Frank Appiah is known as an Ashanti--

Ghanaian pioneer in computer engineering,

software engineering and computing. He currently

holds academic/teacher positions at King's College

London and Kwame Nkrumah University since

2010. Appiah is a professor of informatics,

software engineering and computer engineering.

Frank Appiah has been holding both academic and

teacher positions after these programmes; Bsc.

(Bachelor in computer engineering, Kwame

Nkrumah University) in 2008, Msc.(software

engineering, King's College London) Level in

2009/10 and PGStd (computer education, post-

graduate student, King's/IEEE) diplomatic level in

2010. He holds several doctorates from King's

College , Kwame Nkrumah University and more.

He has about 20+ publication to his credit and

published about 7 books.

Table of Contents

Index of Tables

CHAPTER 1

INTRODUCTION.

1 Technology: A natural and evolutionary processes

Technology[2] is a system based on the application of knowledge, manifested in physical objects and organizational forms, for the attainment of specific goals. But compensating for this physical weakness is an intelligence that is the ultimate source of technology. The natural process of technology is the characteristics of species in a natural world that gives power to sophisticated lives in nature. The ecosystem of living creatures is a broad level of sophistication and the power over nature is the power to destroy or enhance not only our own lives but also the very existence of

every living creature on the planet. The question

that one has to explore in understanding the

process of technological change or aspect of

technology process in our life is totally enmeshed

in the following:

- Are we trying enhance aspect of life with technology?

- Is there an artificial construct for our natural consequence?

- Is there a way to separate from the natural process of doing things?

- Are we attempting to thwart nature?

- Is there a way to escape our senses with technology?

- Is there a way to address concerns of non-biblical soundness?

- Is there a field to enrich ourselves from technology?

The characteristics of specie are acquired from the story of our everyday lives. Here, a story from [1],

George and his companion spent much of their time ranging out cross a plain near their most recent camp site in search for food. They came across a small creature which lives on a wide, flat plain in Africa. The creature was unusually hungry. George is the name for Paul's purpose. They met some hunters who are likely to devour the near-by fruit before they can get their hands to it. They were to keep in touch with time with some insects or reptiles around the site. George and his fellows needed fuel to satisfy their internal furnace. With the fire, a desire for a great kill for a supply of protein is evenly dwell upon. Today they were near a high rock carapace to the east, though they had no concept of direction in those terms. It was merely "the high place over there, where the sun rises." George was scouting ahead of the pack, a chore he seemed to relish. He enjoy the prospect of being the first to sight a potential prey. The band was in the narrow passage that led into the center of the mound of rocks, close to where a fellow hunter had perished only days before at the hands of another predator, a huge cat creature with claws to tear at the throat and jaws to sink deeply into the flesh and break the victim's neck. The memory of the danger remained solidly in his mind. Shorted for clarity.

George used the combination of events to

grasp meaning to his advantage in preserving his life. The fact that he was capable of doing that stems from the fact that nature had supplied him with a brain capable of making those vital connections. Without this ability, he would most likely have perished. The facility to reason, which is what we are really dealing with here, is a *survival trait,* that is a characteristic of the organism that enhances the creature's chances of survival in a hostile environment. This is part of what nature does, and it is part of what has come to be known as *evolutionary theory.* Evolutionary theory[1] is a hot subject in recent times. Evolutionary theory is a useful theory that tends to support the bulk of scientific evidence and one that is increasing in so-

phistication and refinement. Evolutionary theory is

a method of explaining real-world phenomena.

The study of technology are the concepts of

survival of the fittest, natural selection,

specialization and adaptability in a society.

1.1 Natural selection and survival of the fittest.

These are passive phenomena, representing

observations of the way in which nature appears to

operate. Basically, natural selection is a

combination or any path along developmental

scale that is possible but only one of that path

exists because of it's superiority as a format that works. Natural selection is a trait occurrence of usefulness of one specie having an edge over the other specie. The edge of one trait over another through time will be the reason of specie trait survival than those without the trait. Natural selection is made to take place in a nature. The natural environment provides the species a means of survival by competing with the environment. With the natural environment, certainly facts become rapidly apparent. Finally, specie s in a natural environment is filled with a natural trait of survival by means of fittest best suited to carry out activities connected with its task environment.

George events in a natural environment that causes

the following processes of determining technology

by natural selection and survival of fittest:

#	Event	Natural Technology
1	George spent time ranging out cross a plain in search for food.	**Food**: This can be a seasonal database that show when, which kinds of foods grows on the plain.
2	George spent time ranging on plain near camp site.	**Car or Plane**: This can help with the fly over in the purpose of less time search in a plain.
3	George spent time ranging on plain near camp site.	**Tent**: This can help in keeping George needs safe in pack and at rest.
4	George still needs time to range out without the attack of the hungry, small creatures.	**Telescope**: This can help in their search by ranging out from a long distance and timing their movement safely in the environment.
5	George came across small and hungry creatures.	**Stone or Weapon**: This can be thrown at them to drive them of their

		path.
6	George can use the small creatures to test for poisonous substances in the fruit.	**Health Tester** : This can help them not to get sick in the process of eating the wild fruits.
7	George and companion needs to satisfy their internal furnace.	**Wood:** This is a traditional means of providing fire for furnace purposes.
8	George needs to keep in touch with the reptiles at the site.	**Metal Cage** : This is metal protection fixed into the water or land in the case of reptile attack.
9	George needs to keep in touch with the insects at the site.	**Net** : This is a square hollow protection around the body from head to feet.
1 0	George needs to hunt and preserve to	**Trap and Roast or Cook**: This will preserve it to keep it's

	survive for the weeks ahead at the site.	natural scent for some time.
1 1	They are near a high rock and had no concept of direction.	**Coal or Muddy Marker |Compass**: They can mark the rock with a symbol to indicate foot-step direction.
1 2	It is merely a high place over there where the sun rises.	**Dry**: The heat from the sun can be used to dry cloths, meat, crops or more whiles at the site. It is a matter of taking it there.
1 3	George has a memory of danger and cannot forget.	**Pleasure of Tree**: The pleasure of sight of trees or flower plants can give a different picture in the memory.

Table 1: Natural Technology

1.2 Evolutionary Process, Specialization and Adaptability.

The evolutionary theory[1] as said in technology is by way of analogy in general. The manipulation of primitive organisms is an example of technology by human operations. The difference between humans and animals lies in the way humankind goes about the evolutionary process. The learning of what works in the organism ecosystem gives us at least a model to do with technology. The concept of specialization carries the process of

natural selection and survival of fittest to a further

process. The best able to survive in a given

environment are those ones who will survive there,

we find that nature will select more and more

succinctly for traits that fit in with the environment

in question. Specialization is the tendency of an

organism, through the lapse of time to

retain/obtain only traits that lend themselves to

survival and to give up or forgot those traits that

are not useful to their end. Economy is a basic law

in nature. Everything is done in way to such as to

maximized efficiency and not waste, which is not

support by nature. Technology is a better mimic of

nature at best in principle. Thus in maintaining the

efficiency of the organism ecosystem, nature gives

us best as it knows it. Specialization technology

fall into two main categories:

1. *Positive Specialization*: This is the ability

   of an organism to survive in a given

   environment by selecting those trait that

   ensure success.

2. *Negative Specialization*: This is the ability

   of an organism to remove traits that do not

   lend themselves to survival in a specific

   environment..

An organism becomes more dependent on the

specific environment as it becomes more

specialized and therefore more efficient at survival

within a given environment. Adaptability is the

way by which human manufactures organism

technology. In order to adapt , humans extends

themselves into the organism's environment

surrounding not restricted by internal change. The

internal combustion engine is used in cars or trains

to give us greater mobility at a rate that surpasses

any. By extending our self from bodies outwardly

into an environment at the same time changing

the environment within which we are operating

suits our purpose. Humankind does evolve. By

instinct, the transmission of information

concerning what is successful and unsuccessful in

a given environment is done in most cases through

the genetic process of DNA coding. George events

in a natural environment that causes the following

processes of determining technology by

specialization and evolutionary processes:

#	Event	Specialization Technology
1	George spent time ranging out cross a plain in search for food.	Database Technology.
2	George spent time ranging on plain near camp site.	Automobile or Aerospace Technology.
3	George spent time ranging on plain near camp site.	Plastic Technology.
4	George still needs time to range out without the attack of the	Optical Technology.

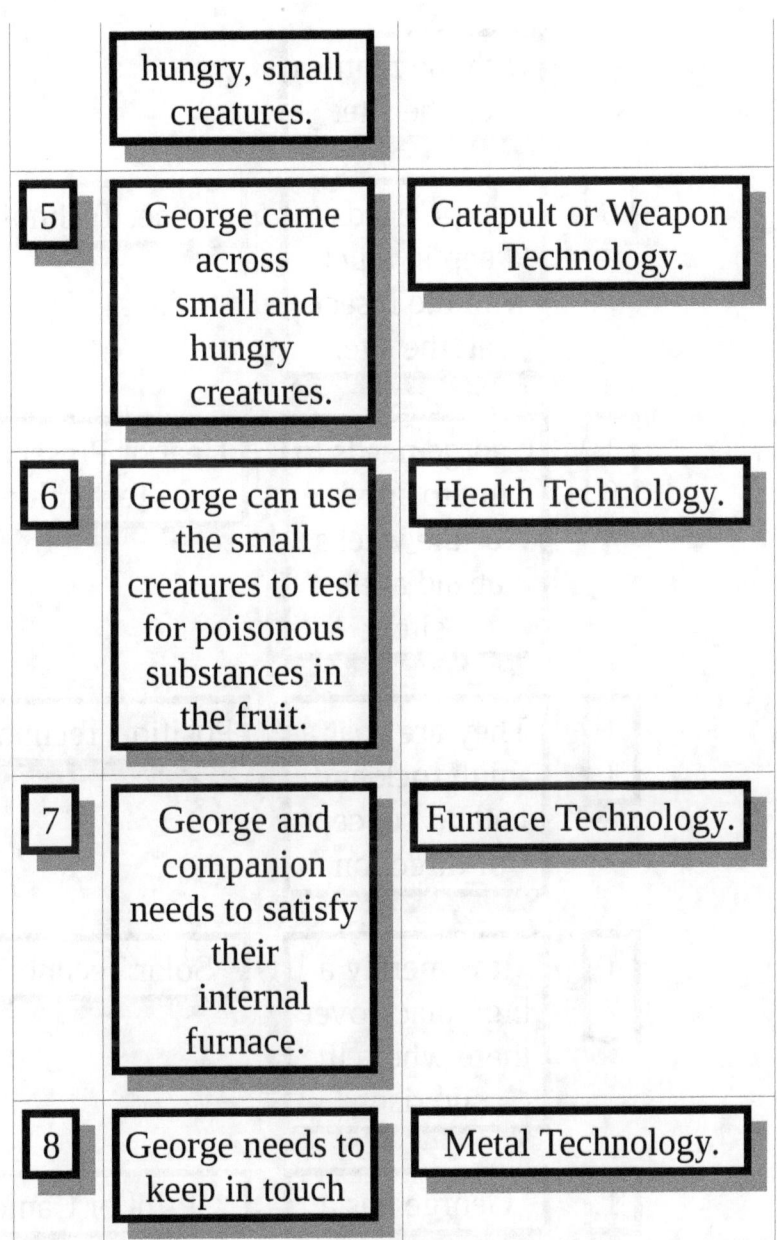

	hungry, small creatures.	
5	George came across small and hungry creatures.	Catapult or Weapon Technology.
6	George can use the small creatures to test for poisonous substances in the fruit.	Health Technology.
7	George and companion needs to satisfy their internal furnace.	Furnace Technology.
8	George needs to keep in touch	Metal Technology.

	with the reptiles at the site.	
9	George needs to keep in touch with the insects at the site.	In-net Technology.
10	George needs to hunt to survive for the weeks ahead at the site.	Heat or Preservative Technology.
11	They are near a high rock and had no concept of direction.	Position Technology.
12	It is merely a high place over there where the sun rises.	Solar Technology.
1	George has a	Print or Camera

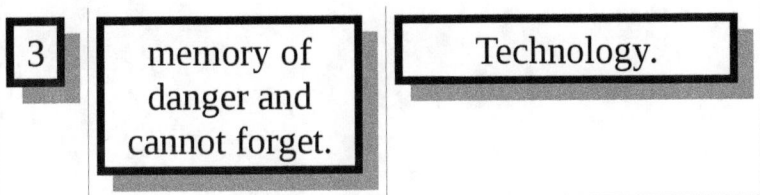

| 3 | memory of danger and cannot forget. | Technology. |

Table 2: Specialization Technology

CHAPTER 2

CREATIVE PROCESS.

2 Creative Element.

There is a need to see things in a new light. The creative process is the critical link in solving a problem from an individual perspective. To be creative, to be able to develop new machines, new forms of technology or new ways of doing things requires some sort of creative element. It requires some one with a creative element to identify or see the hidden ones in a big problem and decompose to bring about a solution. The capacity to equally make a cognitive switch in our approach when the needs arises is the creative process. A large tried-and-true samples can be cling and sorted by a creative person based on natural and evolutionary

facts. Creativity is the way a technology is created

for the usefulness of humanity. Creativity scraps

off the boring way of doing things and place us

into a new perspective necessary for things to

evolve for the better. Creative element is to hold a

problem in a different context so as to narrow the

things or ways of doing a thing. An investigation

of an unaltered contextual view of reality is an

extremely useful thing for humanity, as has been

asserted[1].

> *Creativity can be defined as the ability to*
> *combine a number of factors to achieve a*
> *solution to a problem or make an artifact*
> *that is both novel and useful.*

The creative work is about arrangement of known facts into new patterns in order to develop new constructs useful in accomplishing what needs to be done. This is in the subject of prior knowledge. Creativity is an experimental experience in natural process. These experiences affect the creative works of a taught person. Creative teams provide the range of possibilities available than in a single person's mind. The capacity for problem solving is an ability to seek solutions of greater degree of creativity.

Technology and creativity can be classified as a innovative means of moving into a golden goal abstraction.

The author's own experience is motivated by the concept of risk and reward. The creative process can be experienced if the risks involved are low and there is a reward for operating that creative works. This is known as "Pleasure Principle" by Freud. It is both national and institutional way of promising a feeling of accomplishment. Some say "if a child wants a problem we show the child red things"- Ashanti Adage and risks act as a deterrent to behaving in a creative manner. A motivation of seeking pleasure means to avoid pain by a human being. The creative process starts on a hierarchy of needs. This is a motivational factor present in the creative process. Abraham Maslow offers us a hierarchy of needs. Through his research, he

developed an inventory of needs that must be

satisfied in humankind if a person is to achieve a

happy, successful and balanced life. Maslow[1]

divides motive for actions into five need

categories:

1. *Physiological motive of needs*: These are

    survival needs. They include most bodily

    needs such as freedom from thirst and

    hunger, the need for shelter, the need to

    continue the species (sexual drives), and

    other bodily functions.

2. *Safety motive of needs*: These are survival

    needs. They represent not only to be sated,

warm and healthy, but also to ensure that these conditions will continue to exist in the future. Safety needs involve security and protection from harm, either emotionally, mentally or physically.

3. Social or Belonging motive of *needs:* It refers to the needs to belong to and be part of a group or society. Human beings are gregarious in nature and require the presence of others for their well-being. This need offers and receives affection, acceptance, friendship and a general feeling of belonging to some group.

4. *Self-esteem motive of needs*: These are

based on the internal and personal feeling

about one's self. Self-respect, autonomy

and independence are considered as the

higher order of needs.

5. *Self-actualization motive of needs*: This is

the actualization of the higher order of

needs that represent a person to strive at

any point in time. It involves the strive for

full potential to fulfils one's highest

aspirations and grow as an individual.

The three main factors to be considered in the

comparison of oriental and occidental approaches

in creative restrictions are :

1. *Cultural Restrictions*: Cultural influences

by both African and Western cultures have

exerted influence on the way technology

develops in this earth. Ashanti cultural

arrangement of technology lies in the bars

of gold. This gold has being used in the

production of jewellery and necklaces. Any

influence to draw these people to other

technological changes will be a drawback

for them. Until now, gold has found it's use

in the nanotechnology. It's foundation of

technological change is experienced now

after 150 years. The cultural adoption of

gold is because of it's shining properties. It

shines like the sun, maybe the sun causes

the formation of gold but on the paradox,

the Northern part of Ghana has more

sunlight than any. This is what I call "Gold

Field". Gold is mined from the fills of God

as believed but the Ashantians. It is an

expensive process of acquisition that is the

reason they decorate their body with it. The

home of ores or metals is in Ghana. Ghana

and Ashantians are much into jewellery

technology. God promise of technological

change in the world has brought a lot of

respect to this cultural group. If it is

believed in the source of metal then Ghana

and Ashantians are golden age of

technology since B.C. The important

inventions of first developed was drawn

from the physical properties of gold. It is

shines like yellow. Possibly the colour of

yellow can be sourced from gold. Guess

what this is. That is the bulb that shines

yellow. If the gold shines like yellow then

the a technological change can arise from

the devising a bulb that shines like yellow.

This is Ashantians creativity. This is

impressive. The inventor or contractor of

machines or devices has sourced much of

these technologies from these ores or

metals. English chemist, Rutherford is one

who did experiments with the metal, Gold.

A regional restriction from the countries of

Ghana can be a big blow to the

advancement of technology in the world.

Ashantians and its members on agreement

to strive this world into much higher

technological change has opened its doors

to greater trade with signing on commodity

markets. This is an open technology

initiatives from the Ashantians and it's

members. There is a cultural restriction

from Africa if a game of ownership is

played from the Western cultures. The

Western European countries, in contrast,

were extremely slow to develop an

inventive tradition, plodding through

centuries of stagnation and extreme

restriction in the availability of knowledge

before the blooming of Western

technology[1]. The belief of colonization

brought much of them into contact with the

Ashantians. The mental belief is growing

into what might have caused this. It is due

to the earlier civilize people of the world

and this eagerness to develop fast.

Egyptian dealers of gold who did not

caused any technology from their part

caused the fall of Ashantians. We did ban

them from our jewellery technology since the last 100 years. It was a trade agreement to invent not to innovate with the Gold metal. They did become a people of civilization via their trades. Ashantians were focusing on their jewellery technology. They still call themselves rich because of the source is a rich gold from a mine. A traditional Ashantian, now has greater say in the politics of technology than ever. Their approach of "don't care methods" to technological change is sometime to think about. Trade, trade, trade and let the cultural restrictions of the world change in a technological way. The

so-called "Blacksmiths" are not just black

but via their skin, they thought of light

before the Bible reason could. By the light

of God , thou shall find healing in his

name. Gold and Bible are lights from

elements of God's promises. Fantes,

member of Ghana, land of ores upon

meeting the Ejisuian of the Ashantians,

signed to the slave-master agreement in the

purpose to invest heavily with their ores in

any case of technological change sorted

from the home of inventions, Ejisuian. The

slave-master agreement was very simple

until the termed colonization of Western
Europeans. The unthinkable was a thing to
avoid in the process of acquisition of these
metals or ores. The toys from these metals
did not improve the lives of these people
especially Fantes. Ashantians did place a
ban for none to use this precious metal
any-more in machine form or other until
the 21st century. The distinct classes of
technological change from the Ashantians
is hailed by the British and their home has
never found not of such people. The metal
places a position of Ashantians to global
citizens of technology and not forgetting

the Fantes. A honorary appeal to the

inventors or creators or contractors is that

bring back all our technologies from the

ores employed in the process. Period!!!!

After all Europe change from much of the

trade of metals or ores and the land (4+

million people) of possibilities, we

Ashantians showed the way.

2. *Linear Thinking*: Linear thinking is about

thoughts that are linear in nature. It moves

through a sequence of of defined

statements that lead to conclusions about

the world. Logical analysis in Western

cultures are bound by statements with their

cause-effect relationship. Each statements

infers the next. Facts are combined into

syllogistic combinations designed to prove

that if one event occurs or if one set of

circumstances exits, then another event or

set of conclusions must logically follow.

Each culture experiences the feeling of

paradox when dealing with the logic of

Westerners and Orientals. One moves step-

by-step from conclusion to conclusion,

each building on the preceding one in a

web of well-founded proofs toward the

ultimate conclusion being sought. For

instance, if you have a ball in an open box

in the palm *then* you move side-to-side. It

will *then* definitely create a sound (ka-ka-

ka-ka) due to falling from one slope-side to

down-side. If these events do occur *then*

the ball is made of metal, rounded and is

placed in a plastic box with about 2.5

inches tall. *Then* the person who slants box

in a slope is stressful or depressed.

Therefore, if the person drop the ball from

box under the conditions, *then* the person

will be seriously depress or stress. It can be

important in psychiatry for those with

sleepiness problems. This is the type of

linear logic experienced in some cultures in

Ghana. Books construction also follows

occidental writing – word-by-word,

sentence-by-sentence, paragraph-by-

paragraph, front-to-back, top-to-bottom,

chapter-by-chapter, and so physically left-

to-right. Every action has a reaction, and

each reaction becomes an action, creating

another reaction.

3. *Philosophical View*: The balance of the

physical reality of a world is to idea as to

name and study. This is to present the

good-evil part of nature or movement

toward the balance of serenity in nature

and humanity. Balance is the key in the

process. The reality of thing in nature will balance if we know of the ideal name of everything in our environment. The opportunity for exposure for ideas arose for new solutions to answer questions and new information, just a love for knowledge. Invention is a purpose of idea patenting. Ghana has inventive culture, organized and highly developed.

2.1 Scientific Method.

Most scientific representations follows five methodologically steps through which investigators can discover the truth of a problem:

1. *Defining the problem*: It is first necessary to have a clear idea of exactly what the problem is in order to choose a direction of investigation and properly formulate later experimental techniques.

2. *Observing the evidence*: An accumulation of information is necessary as an initial step in investigation to ensure that all previous work on understands the nature of the problem under investigation.

3. *Forming the hypothesis*: The investigator uses intuition and logical thinking to discover perceived patterns of behaviour

from the preliminary data and draw

tentative conclusions from those patterns.

This is a primary creative step by which a

new way of viewing the nature of the

reality under investigation takes place.

4. *Experimenting*: A method of testing is

designed to either validate or deny the truth

of the hypothesis. The design of the

experiment itself is crucially creative

process. The nature of experiment will

automatically exclude and include what

can be discovered by its results.

5. *Formalizing the theory*: The results of the

testing of the original hypothesis and

conclusions are made available to the

public for scrutiny of the investigate

community as a whole and to use to

further investigate the subject.

CHAPTER 3

CONCEPTION OF

FIELDS.

2 Fields Conception.   *God's Fields*

The conception of field has various definitions. Let's look at the farming field first, then seen field and lastly, the heat field.

2.1 Farming Field.

Lets look at the definition of field concerning area of farming. In this sense, a field is an area for farming. A pear plant can be cultivated and one will come to no short of pear gift. One will stay on life with plant thing and get enriched in all things. Then it is to say that the area for farming is God's field. The area for farming is the soil ground for cultivation and all knowledge about soil can be

cultivated. One who is a worker is never

concerned with the pear reward yielded from his

labour. The fellow worker will be rewarded with

all knowledge about soil, water and plants. In the

sense of contention, the household has stayed on

life with the pear plants and all plant things but

how does the whole world go about life with

contentions. This can be answered by a class 12

person.

2.2 Seen Field.

A field is an area that can be seen. One's work will be build on the foundations of fields with gold, silver, precious stones, wood, hay and straw. One will build a global contention-aware field and the world will stay on a life without contentions. Fellow workers can build on the foundations of fields with gold, silver etc. and create equipment to bring something to the world. The naturalist can build a piece of telescope to see the stars and more if one wants commission. Television is fielded that can be seen and one can cultivate a foundation to stay on life with all things.

2.3 Heat Fields.

A field is an area with gas, coal and more. Those who writes their assays in the dark will need each work to become clear for the day. A clear day is revealed by fire from the burning of gas, coal and more. A class 13 person can answer to what is after-yield? In this age, households cannot stay off on pear contentions so the yield needs to be distributed and this is via transportation like vehicles, trains, ships, planes etc. These yield transports all use or some use gas, coal or more. The yield of each worker will be declared in each day after transportation at a particular port. The test of an after-yield will depend on the speed of

firing of the transport. Secondly, if anyone builds

on the after-yield by fire then can be rewarded

from the further preparation like the cooking

methods – Roasting, Boiling, Steaming and the

raw storage.

CHAPTER 4

SUMMARY.

3 Summary.

This chapter summarizes the introduction notes on technology from the perspective of natural and evolutionary processes. The natural processes of technology here did looked at natural selection and survival of fittest. The evolutionary processes of technology are based on the conception of specialization and conception of adaptability. The determining technology of both natural and evolutionary processes are tabulated to address the concerns of the fictional notes of George. The conception of fields are looked at in this studies. In this sense, a field which is an area for farming is noted. The area for farming is the

soil ground for cultivation and all knowledge

about soil can be cultivated. The seen field is

noted as an area that can be seen. The naturalist

can build a piece of telescope to see the stars and

more. In that contention sense, the household has

stayed on life with the plants and all plant things.

Recently at my home, a couple of straw plants

grow within the fence of the walls. I decided to

uproot the straw and after tasted it. I found out that

it contains some glucose. I wonder!!! Heat field is

an area with gas, coal and more. By heat fields, a

clear day is revealed by fire from the burning of

gas, coal and more.

References:

1. Paul A. Alcorn, Technology, Society and Culture. Person Custom Publishing.2002.

2. Linda Hjorth S., Barbara Eichler A. , Ahmed Khan S. and John Morello A. (2002) Technology, Society and Culture, Pearson Custom Publishing.

3. Frank Appiah* (Nsiah). Pounding Task: A task-oriented design1 , series 1 of wusair. Academia.edu.

4. Frank Appiah* (Nsiah). Pounding Task: A task- oriented design 2 , series 2 of wusair. Academia.edu.

5. Frank Appiah* (Nsiah). Pounding Task: A

task- oriented design 3 , series 3 of wusair. Academia.edu.

6. Arthur W. Burks, editor. Theory of Self-Reproducing Automata, University of Illinois Press, Urbana, (1966).

7. Yoshihide Igarashi, Tom Altman, Mariko Funada and Barbara Kamiyama. Computing: A Historical and Technical Perspective. CRC Press, Taylor &Francis Group (2014).

8. Micheal Sipser. Introduction to the Theory of Computation, PWS Publishing Company, (1997).

9. Barry McMullins. John von Neumann and

the Evolutionary Growth of Complexity:

Looking Backwards, Looking Forwards,

MIT Press (2000).

10. Frank Appiah* (Nsiah). A correct

specification development in engineering

design and systems,(2000). Academia.edu.

11. Frank Appiah* (Nsiah). Elements of theory

of a cat control knowledge(2018).

Academia.edu.

12. Frank Appiah* (Nsiah). Elements of

simulation model of cat causations-

effections. Journal of Simulation, CRC

Group, (2018).

Alphabetical Index